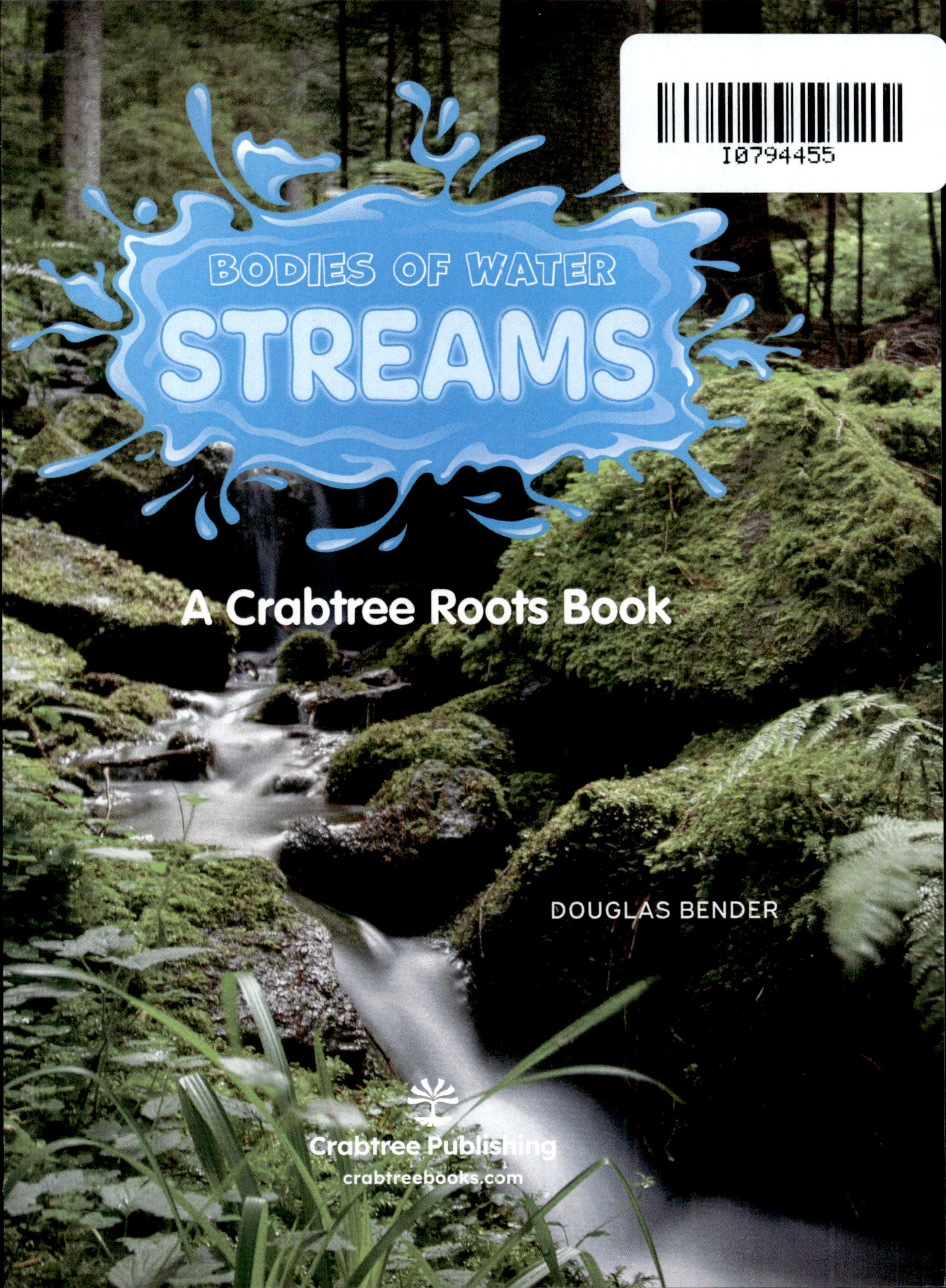

BODIES OF WATER

STREAMS

A Crabtree Roots Book

DOUGLAS BENDER

Crabtree Publishing
crabtreebooks.com

School-to-Home Support for Caregivers and Teachers

This book helps children grow by letting them practice reading. Here are a few guiding questions to help the reader with building his or her comprehension skills. Possible answers appear here in red.

Before Reading:

- What do I think this book is about?
    - *I think this book is about streams and what they look like.*
    - *I think this book is about how streams are made.*
- What do I want to learn about this topic?
    - *I want to learn where streams come from.*
    - *I want to learn what size a stream is.*

During Reading:

- I wonder why...
    - *I wonder why you can drink water from streams.*
    - *I wonder why lots of snow and rain make streams.*
- What have I learned so far?
    - *I have learned that streams have rocks.*
    - *I have learned that streams are small bodies of water.*

After Reading:

- What details did I learn about this topic?
    - *I have learned that streams are found in places where it rains or snows.*
    - *I have learned that water in streams moves around rocks.*
- Read the book again and look for the vocabulary words.
    - *I see the word* ***stream*** *on page 3 and the word* ***rocks*** *on page 7. The other vocabulary word is found on page 14.*

This is a **stream**.

Most streams are little.

There are many **rocks** in streams.

Some streams come from snow.

Some streams come from rain.

Many streams have water you can **drink**!

Word List

Sight Words

a
are
can
come
from
have
is
many
there
this
rain
snow
some
water

Words to Know

drink

rocks

stream

31 Words

This is a **stream**.

Most streams are little.

There are many **rocks** in streams.

Some streams come from snow.

Some streams come from rain.

Many streams have water you can **drink**!

BODIES OF WATER

STREAMS

Crabtree Publishing

crabtreebooks.com 800-387-7650

 In Canada: We acknowledge the financial support of the Government of Canada through the Canada Book Fund for our publishing activities.

Hardcover	978-1-4271-5595-5
Paperback	978-1-4271-5601-3
Ebook (pdf)	978-1-4271-3378-6
Epub	978-1-4271-3438-7
Read-along	978-1-4271-5619-8
Audio book	978-1-4271-5625-9

Published in Canada
Crabtree Publishing
616 Welland Avenue
St. Catharines, Ontario
L2M 5V6

Published in the United States
Crabtree Publishing
347 Fifth Avenue
Suite 1402-145
New York, NY 10016

Written by: Douglas Bender
Designed by: Rhea Wallace
Series Development: James Earley
Proofreader: Janine Deschenes
Educational Consultant: Marie Lemke M.Ed.
Photographs: Shutterstock: irin-k: cover; normaniac: p. 1; pillster: p. 3, 14; Gwoeii: p. 5; Chase Clausen: p. 6, 14; Irina Boldina: p. 8-9; andreiuc88: p. 11; Anton Watman: p. 13, 14

Printed in Canada/042024/CPC20240409

Library and Archives Canada Cataloguing in Publication
Title: Streams / Douglas Bender.
Names: Bender, Douglas, 1992- author.
Description: Series statement: Bodies of water | "A Crabtree roots book".
Identifiers: Canadiana (print) 20210190205 | Canadiana (ebook) 20210190221 | ISBN 9781427155955 (hardcover) | ISBN 9781427156013 (softcover) | ISBN 9781427133786 (HTML) | ISBN 9781427134387 (EPUB) | ISBN 9781427156198 (read-along ebook)
Subjects: LCSH: Rivers—Juvenile literature.
Classification: LCC GB1203.8 .B46 2022 | DDC j551.48/3—dc23

Library of Congress Cataloging-in-Publication Data
Names: Bender, Douglas, 1992- author.
Title: Streams / Douglas Bender.
Description: New York, NY : Crabtree Publishing Company, [2022] | Series: Bodies of water - a Crabtree roots book | Includes index.
Identifiers: LCCN 2021017118 (print) | LCCN 2021017119 (ebook) | ISBN 9781427155955 (hardcover) | ISBN 9781427156013 (paperback) | ISBN 9781427133786 (ebook) | ISBN 9781427134387 (epub) | ISBN 9781427156198
Subjects: LCSH: Stream ecology--Juvenile literature.
Classification: LCC QH541.5.S7 B46 2022 (print) | LCC QH541.5.S7 (ebook) | DDC 577.6/4--dc23
LC record available at https://lccn.loc.gov/2021017118
LC ebook record available at https://lccn.loc.gov/2021017119